SURVEYING M. MADE S

MW00916097

An original Book by

Jim Crume P.L.S., M.S., CFedS

Co-Authors
Cindy Crume
Bridget Crume
Troy Ray R.L.S.
Mark Sandwick L.S.I.T.

PRINTED EDITION

*****

PUBLISHED BY:

Jim Crume P.L.S., M.S., CFedS

Inverse Between Rectangular Coordinates

Book 3 of this Math-Series

First publication: November, 2013

Printed by CreateSpace

Available on Kindle and other devices

TERMS AND CONDITIONS

The content of the pages of this book is for your general information and use only. It is subject to change without notice.

Neither we nor any third parties provide any warranty or guarantee as to the accuracy, timeliness, performance, completeness or suitability of the information and materials found or offered in this book for any particular purpose. You acknowledge that such information and materials may contain inaccuracies or errors and we expressly exclude liability for any such inaccuracies or errors to the fullest extent permitted by law.

Your use of any information or materials in this book is entirely at your own risk, for which we shall not be liable. It shall be your own responsibility to ensure that any products, services or information available in this book meet your specific requirements.

This book may not be further reproduced or circulated in any form, including email. Any reproduction or editing by any means mechanical or electronic without the explicit written permission of Jim Crume is expressly prohibited.

Table of Contents

INTRODUCTION

Straight forward Step-by-Step instructions.

This book is just one part in a series of digital and printed editions on Surveying Mathematics Made Simple. The subject matter in this book will utilize the methods and formulas that are covered in the books that precede it. If you have not read the preceding books, you are encouraged to review a copy before proceeding forward with this book.

For a list of books in this series, please visit:

http://www.cc4w.net/ebooks.html

Prerequisites for this book:

A basic knowledge of geometry, algebra and trigonometry is required for the explanations shown in this book.

Book 1 - **Bearings and Azimuths** - How to add bearings and angles, subtract between bearings, convert from degrees-minutes-seconds to decimal degrees, convert from decimal degrees to degrees-minutes-seconds, convert from bearings to azimuths and convert from azimuths to bearings.

Book 2 - **Create Rectangular Coordinates** - How to calculate the northing and easting of an end point given the coordinates of the beginning point, bearing and distance of a line.

Definition: Inverse (a.k.a. Bearing and Distance) is the computation for the horizontal distance and bearing of a line from the coordinates of its endpoints.

INVERSE BETWEEN RECTANGULAR COORDINATES

It is important to keep an eye on the algebraic sign for the Lat and Dep. This is an indicator as to the quadrant that the bearing is located.

NE Bearing: Lat and Dep will both be positive

SE Bearing: Lat will be negative and Dep will be positive

SW Bearing: Lat and Dep will both be negative

NW Bearing: Lat will be positive and Dep will be negative

Example 1:

Figure 1 shows a straight line between two coordinate pairs of which the bearing and distance is desired. To calculate the bearing and distance, you must first determine the Latitude (Lat) and Departure (Dep) between the two known points.

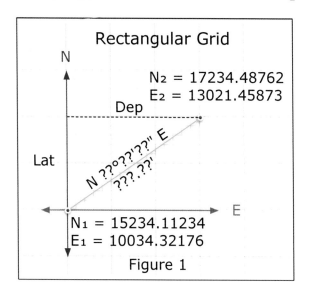

Figure 1

Use the following formulas to calculate the Lat, Dep, Distance and Bearing between known Points N_1E_1 and N_2E_2.

Formulas:

Lat = N_2 - N_1

Dep = E_2 - E_1

Distance = $\sqrt{(Lat^2 + Dep^2)}$ [Pythagorean theorem]

Bearing = Arctan(Dep / Lat) [Arctangent is also known as Tan^-1]

Example 1 - Figure 1 solution:

Lat = 17234.48762 - 15234.11234 = **2000.37528**

Dep = 13021.45873 - 10034.32176 = **2987.13697**

Distance = $\sqrt{(2000.37528^2 + 2987.13697^2)}$ = **3595.06447**

Bearing = Arctan(2987.13697 / 2000.37528

Bearing = 56.19124098° = **N 56°11'28" E**

NOTES

Example 2:

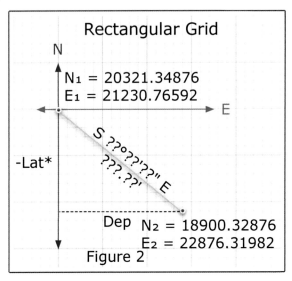

Rectangular Grid

N

N_1 = 20321.34876
E_1 = 21230.76592

E

-Lat*

S ??°??'??" E

Dep N_2 = 18900.32876
E_2 = 22876.31982
Figure 2

* Note the negative algebraic sign

Example 2 - Figure 2 solution:

Lat = 18900.32876 - 20321.34876 = **-1421.02000**

Dep = 22876.31982 - 21230.76592 = **1645.55390**

Distance = $\sqrt{(1421.02000^2 + 1645.55390^2)}$ = **2174.19996**

Bearing = Arctan(1645.55390/ -1421.02000)

Bearing = -49.18771203° = **S 49°11'16" E**

Note: The negative Lat indicates that the bearing is in the SE quadrant.

Examples 1 & 2 walked you through the steps to inverse between two rectangular coordinate pairs. Now that you have a couple of examples to follow, try solving Examples 3 & 4, then review the solutions at the end of the book to see how you did.

NOTES

Example 3:

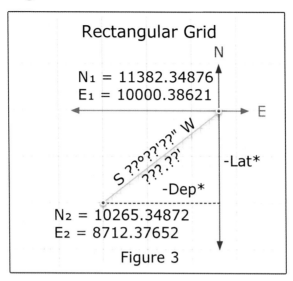

Rectangular Grid

$N_1 = 11382.34876$
$E_1 = 10000.38621$

S ??°??'??" W

-Lat*

-Dep*

$N_2 = 10265.34872$
$E_2 = 8712.37652$

Figure 3

* Note the negative algebraic sign

Figure 3 solution:

Lat = ??? - ??? = **???**

Dep = ??? - ??? = **???**

Distance = $\sqrt{(???^2 + ????^2)}$ = **???**

Bearing = Arctan(???/ ???)

Bearing = ???° = **S ??°??'??" W**

The solution is towards the end of the book.

NOTES

Example 4:

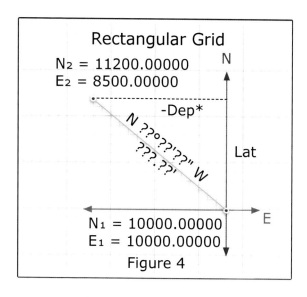

Figure 4

* Note the negative algebraic sign

Figure 4 solution:

Lat = ??? - ??? = **???**

Dep = ??? - ??? = **???**

Distance = $\sqrt{(???^2 + ????^2)}$ = **???**

Bearing = Arctan(???/ ???)

Bearing = ???^{0} = **N ??^{0}??'??" W**

The solution is towards the end of the book.

NOTES

SOLUTIONS TO EXAMPLES 3 & 4

Example 3 - Figure 3 solution:

Lat = 10265.34872 - 11382.34876 = **-1117.00004**

Dep = 8712.37652 - 10000.38621 = **-1288.00969**

Distance = $\sqrt{(-1117.00004^2 + -1288.00969^2)}$ = **1704.89239**

Bearing = Arctan(-1288.00969/ -1117.00004)

Bearing = 49.0672051$6^0$ = **S 49°04'02" W**

Example 4 - Figure 4 solution:

Lat = 11200.00000 - 10000.00000 = **1200.00000**

Dep = 8500.00000 - 10000.00000 = **-1500.00000**

Distance = $\sqrt{(1200.00000^2 + -1500.00000^2)}$ = **1920.93727**

Bearing = Arctan(-1500.00000/ 1200.00000)

Bearing = -51.3401917$5^0$ = **N 51°20'25" W**

Note: Rounding error is dependent upon the number of decimal places that are utilized. It is recommended that at least 5 decimal places or more be used for all calculations then round the final answer as needed.

NOTES

ABOUT THE AUTHOR
Jim Crume P.L.S., M.S., CFedS

My land surveying career began several decades ago while attending Albuquerque Technical Vocational Institute in New Mexico and has traversed many states such as Alaska, Arizona, Utah and Wyoming. I am a Professional Land Surveyor in Arizona, Utah and Wyoming. I am an appointed United States Mineral Surveyor and a Bureau of Land Management (BLM) Certified Federal Surveyor. I have many years of computer programming experience related to surveying.

This book is dedicated to the many individuals that have helped shape my career. Especially my wife Cindy. She has been my biggest supporter. She has been my instrument person, accountant, advisor and my best friend. Without her, I would not be the professional I am today. Cindy, thank you very much.

Other titles by this author:

http://www.cc4w.net/ebooks.html

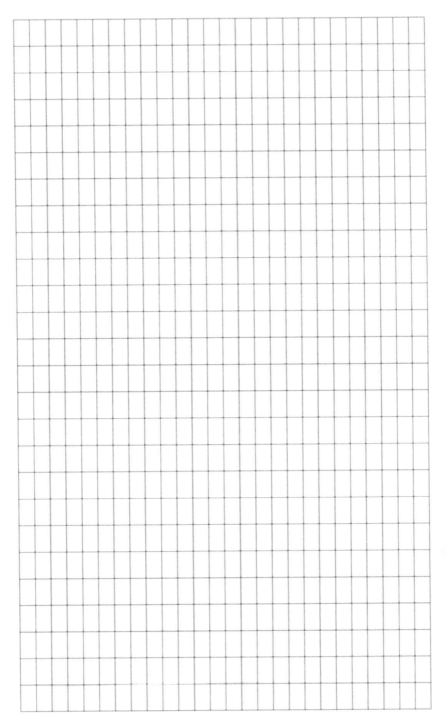

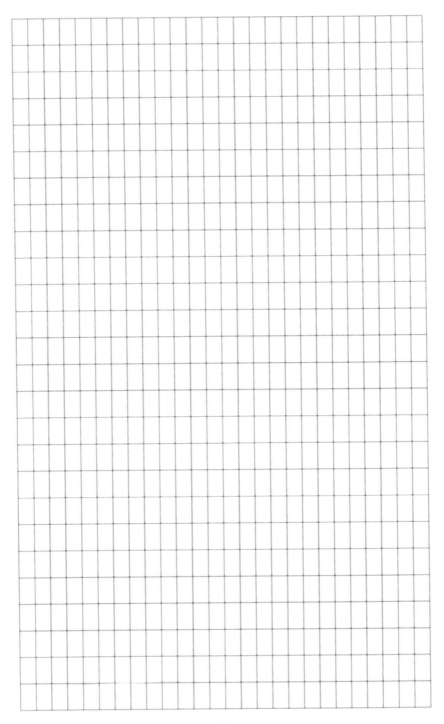

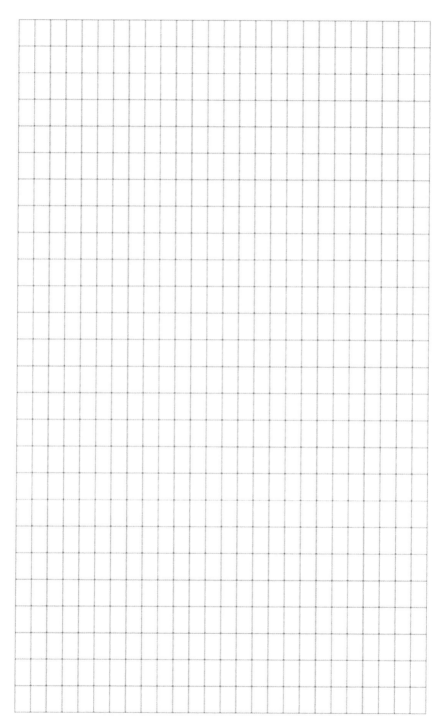

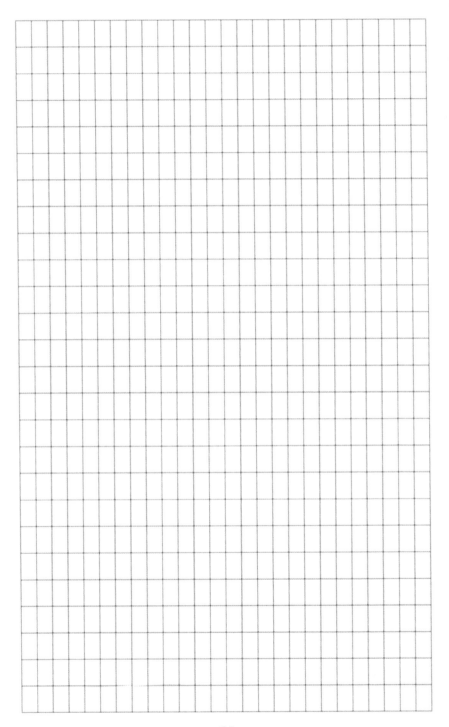

Made in the USA
Columbia, SC
03 March 2020